SHORT TIMERS GUIDEBOOK

By Edward Temple

Dedicated to my wife
Pamela
to whom I'll always be a lifer.

In Memory of
the Desert Shield and Desert Storm warriors
who paid the ultimate price for freedom
during the liberation of Kuwait

Blessed are thc peacemakers,
for they shall be called the sons of God.
Matthew 5:9

CONTENTS

To Fred,

Best wishes as you begin your next career.

Remember all the lessons taught in AARP class.

Semper Fidelis

Edward [illegible] USMC

GySgt

Chapter 1

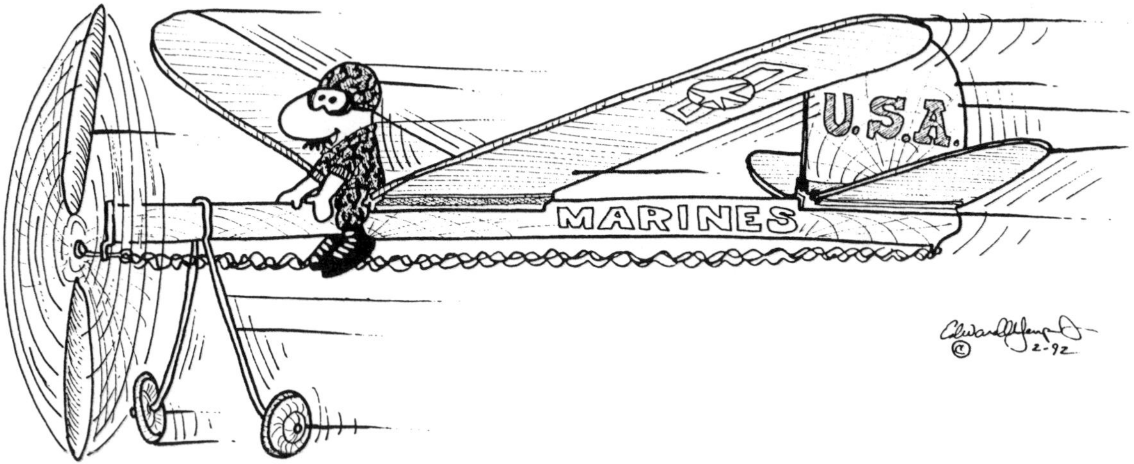

I'M SO SHORT, I CAN RIDE A WIND-UP RUBBERBAND AIRPLANE.

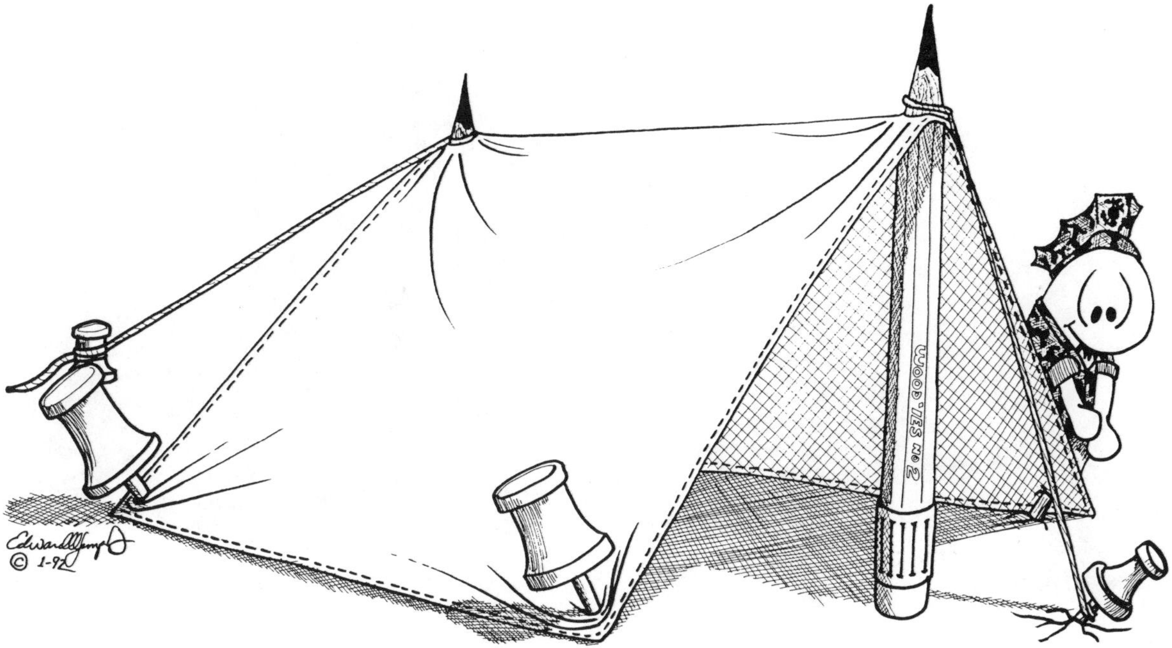

I'M SO SHORT, I CAN USE A HANDKERCHIEF FOR A TENT.

I'M SO SHORT, I HAVE TO PULL DOWN MY SOCKS TO SEE.

I'M SO SHORT, A 100 YARD DASH CAN LAST 28 MINUTES.

I'M SO SHORT, I CAN PLAY HANDBALL AGAINST A CURB.

I'M SO SHORT, I CAN GO WHITE WATER RAFTING ON A LEAF.

I'M SO SHORT, I CAN USE A THUMBTACK AS AN UMBRELLA.

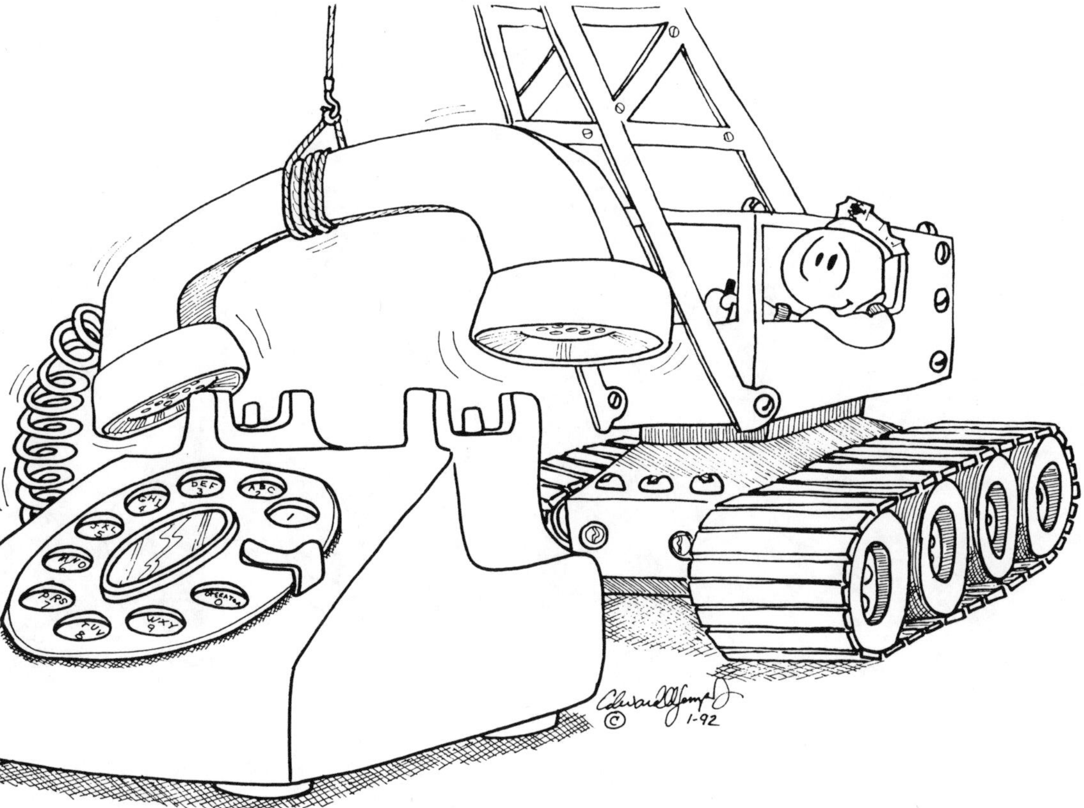

I'M SO SHORT, I NEED A TOY CRANE TO ANSWER THE PHONE.

I'M SO SHORT, THAT GRAINS OF SAND LOOK LIKE BOULDERS.

I'M SO SHORT, GETTING A HAIRCUT IS A LIFE THREATENING EXPERIENCE

I'M SO SHORT, I BUY MY UNIFORMS IN A TOY STORE.

I'M SO SHORT, I CAN RAPPEL OFF OF A DIME.

I'M SO SHORT, A SAND CASTLE LOOKS LIKE THE REAL THING.

I'M SO SHORT, I CAN DO PULL-UPS ON A STAPLE.

Chapter 2

PROS: THEY CAN'T PUT YOU IN JAIL.

CONS: THEY STAPLE YOU TO THE FLOOR INSTEAD.

PROS: ONE LIFE JACKET CAN SAVE TWELVE SHORT TIMERS.

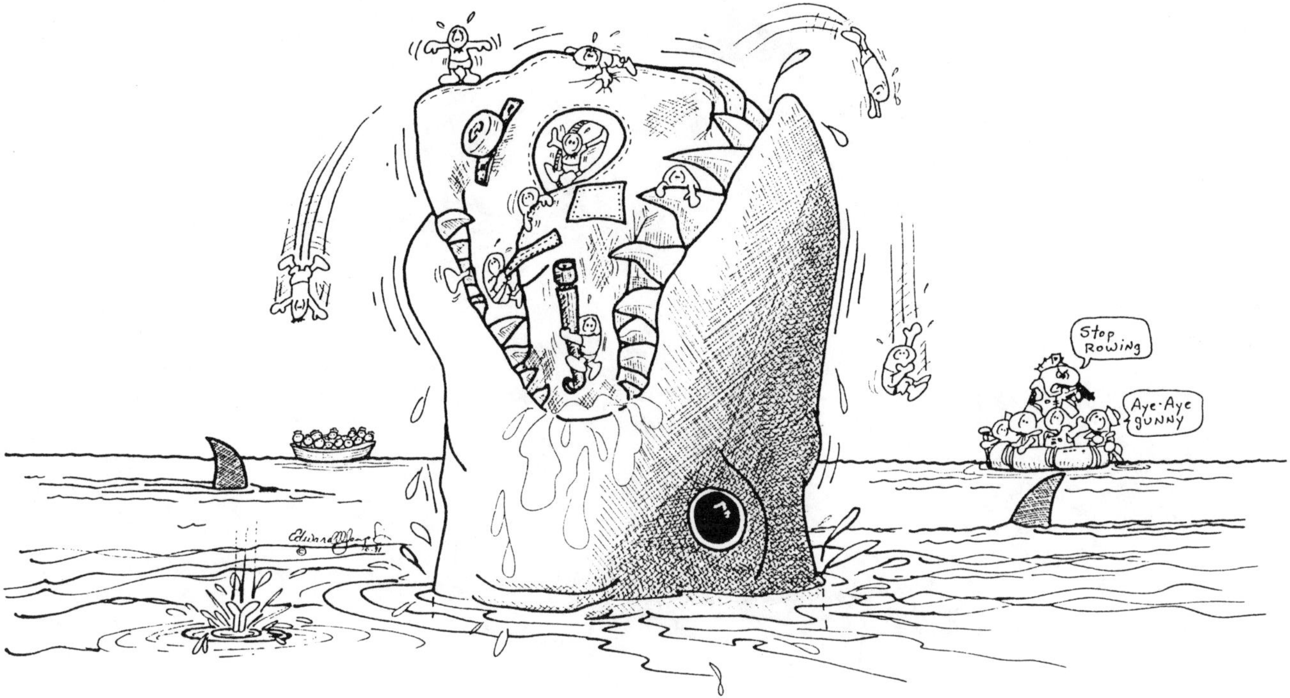

CONS: IT ONLY TAKES ONE BITE TO EAT ALL OF US.

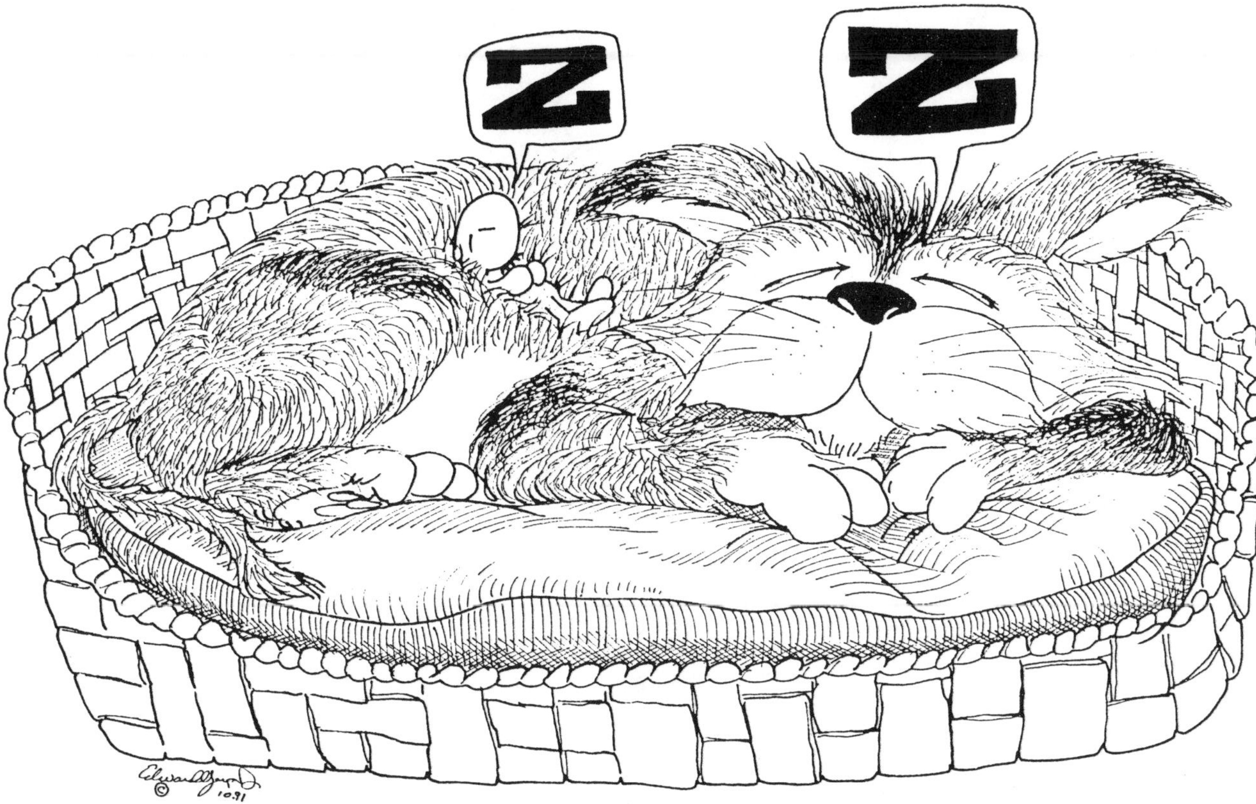

PROS: A CAT CAN MAKE A GREAT BED.

CONS: CLEANING THE LITTER BOX.

PROS: YOU CAN ADMIRE NATURE UP CLOSE.

CONS: JUST DON'T GET TOO CLOSE.

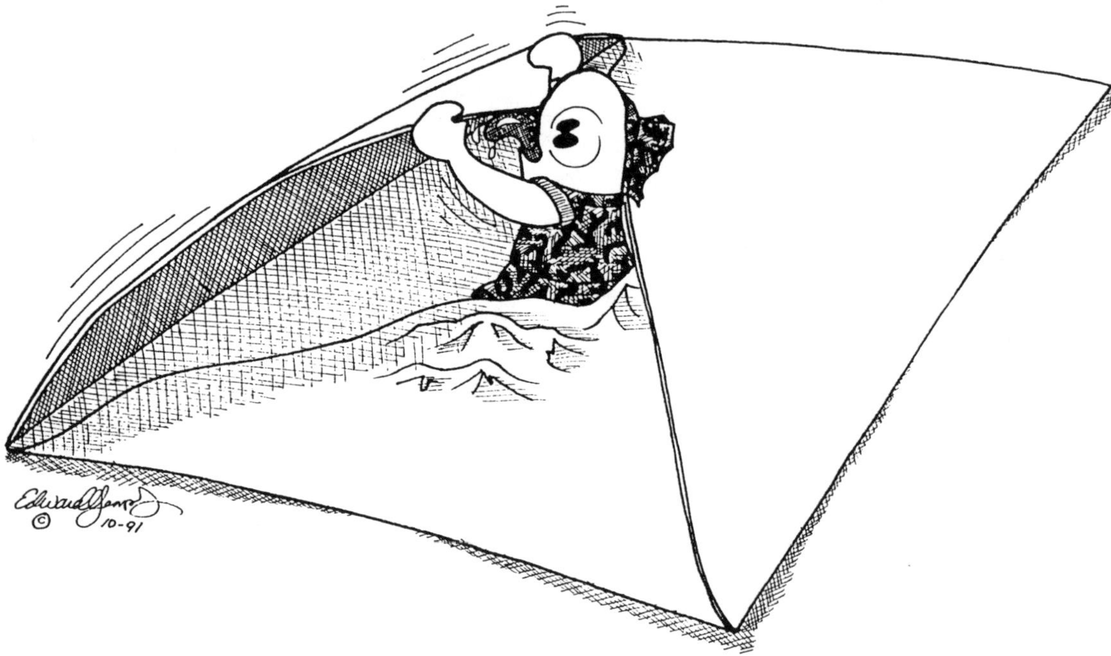

PROS: I CAN MAIL MYSELF HOME.

CONS: GETTING STAMPED.

PROS: YOU CAN FIND SHADE ALMOST ANYWHERE.

CONS: SO CAN ANYTHING ELSE.

PROS: YOU CAN HAVE A 'BIG' BIRTHDAY PARTY WITH ONE CUPCAKE.

CONS: IT'S DEADLY BLOWING OUT THE CANDLE.

Chapter 3

SIX YEARS OF ACTIVE DUTY. NAVY ACHIEVEMENT MEDAL, PRESIDENTIAL UNIT CITATION, NAVAL UNIT CITATION, AND TWO MERITORIOUS UNIT CITATIONS. TWO GOOD CONDUCT AWARDS, 27 MERITORIOUS MASTS, 98 LETTERS OF APPRECIATION, WITH 39 CERTIFICATES OF COMMENDATIONS. 300 P.F.T.'er TWENTY TIMES. YOU WERE AN HONOR MAN OUT OF BOOT CAMP. RIFLE AND PISTOL EXPERT SIX YEARS IN A ROW. COMPLETED 159 M.C.I. COURSES IN ONE YEAR. DURING LEAVE YOU RECRUITED 27 NEW MARINES AND 17 NEW OFFICERS. NEVER HAD OFFICE HOURS OR A PAGE ELEVEN ENTRY.

I DON'T UNDERSTAND, WHY DON'T YOU WANT TO RE-ENLIST , PRIVATE?

SPIT SHINES

U.S.M.C.
SEMPER FIDELIS

DAMN SHORT TIMERS

HOW TO SEND SHORT TIMERS INTO COMBAT

LAST VISIT TO THE CAREER PLANNER.

CUTTING GRASS

HEY, WHO'S GOING TO THROW THIS THING ?

Did you hear a click ?

MOUNTAIN CLIMBING

SHORT TIMERS DRIVING A CAR.

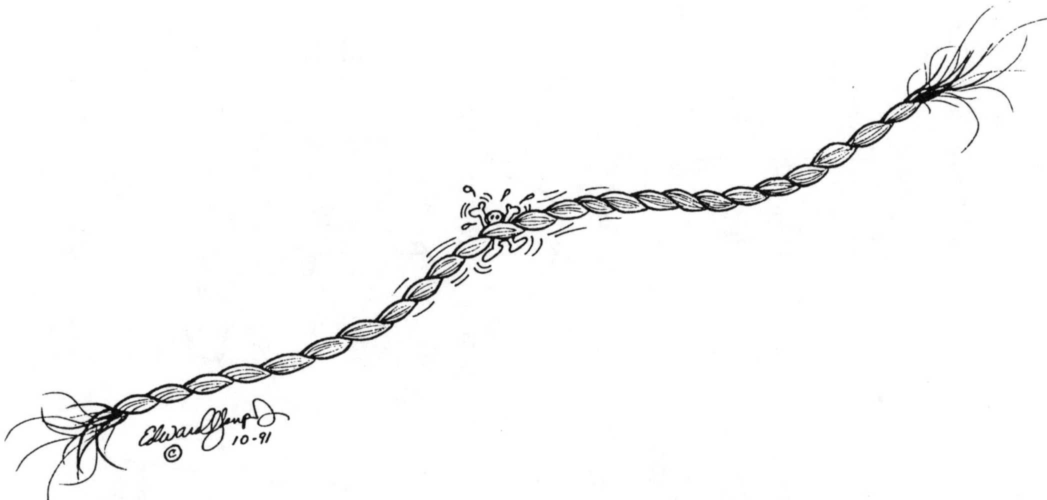

IRISH PENNANT ON A SHORT TIMER.

SHORT TIMERS STEW

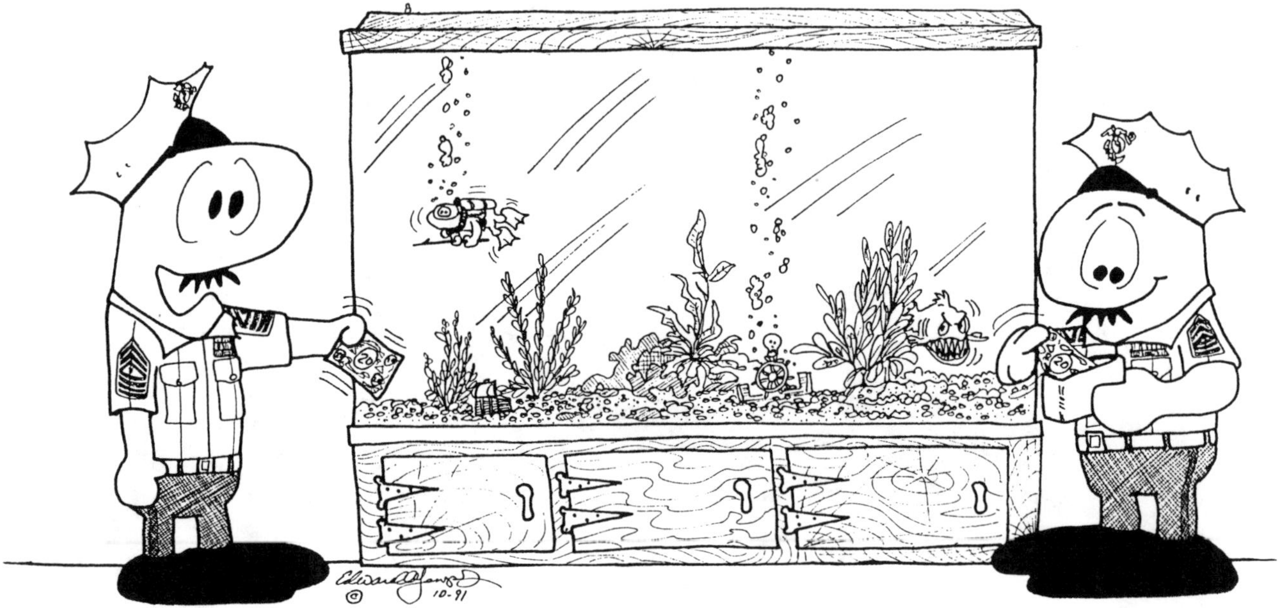

I'LL BET TWENTY BUCKS ON THE PIRHANA.

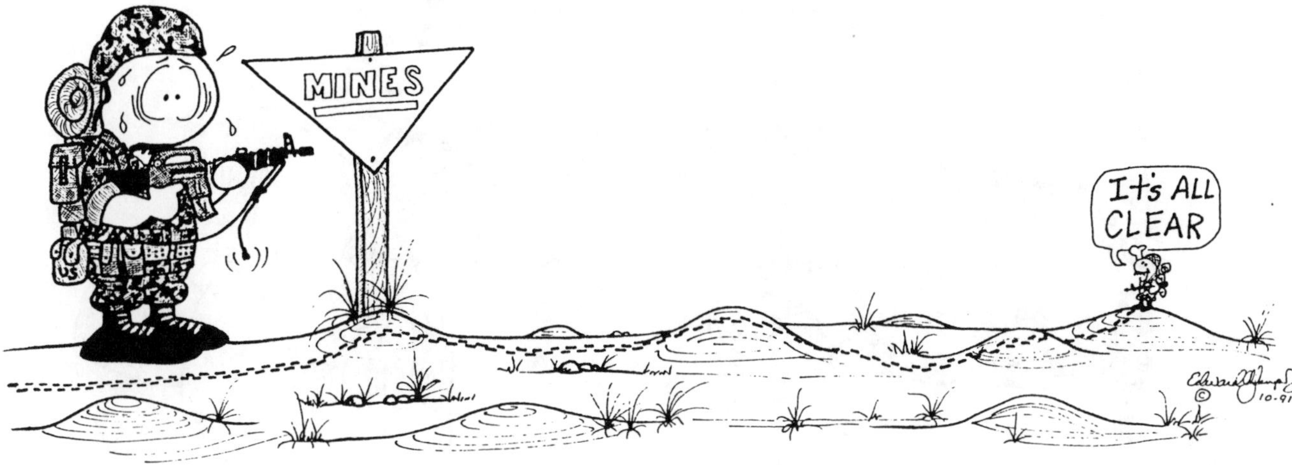
MINES
It's ALL CLEAR
10-91

TOP
BOY
-R-
DEE
B-B-QUE
SAUCE

USMC

SHORT TIMER CAUGHT IN THE RAIN.

THANKS DAD, WHEN CAN I UNWRAP THEM ?

TARGET PRACTICE

I ALWAYS WANTED TO DO THIS!

SHORT TIMERS HEAVEN

YOU CAN START TERMINAL LEAVE ON.....

-370
SEE, I TOLD YA´
10-91

SNACK PACK

12
GUTS and GLORY
HERO
MARINES
GUNG-HO
PLOP
WAR and PEACE

HERE, HOLD THIS PIN OVER YOUR HEAD.

SHORT TIMER SKIING

5:41
ON
OFF
APT. FOR
RENT
FOR
RENT
12:37
© 2-92

©
10-91

THERE'S NO BETTER WAY TO SPEND A SATURDAY MORNING.

THANKSGIVING

LOOKS CLEAN TO ME SERGEANT MAJOR.

A "REAL" GAME OF CHESS.

JUST TWO MORE MINUTES, GENERAL.

HOOKS
GLUE
WOMEN MARINES
CHRISTMAS
DECORATIONS
GARLAND
GLASS
BALLS
USMC
© 5-92

I LOVE THE WAY YOUR DAD LANDS HIS PLANES.

FIRE FOUR

THAT IS THE SMALLEST BANDAID WE HAVE.
© 1-92

FIDO FOOD
CHESTY

WHAT DO YOU DO WITH THEM
WHEN THE BATTERIES RUN OUT?
BATTERY
RECHARGER
PORT-O-LITE
3-91

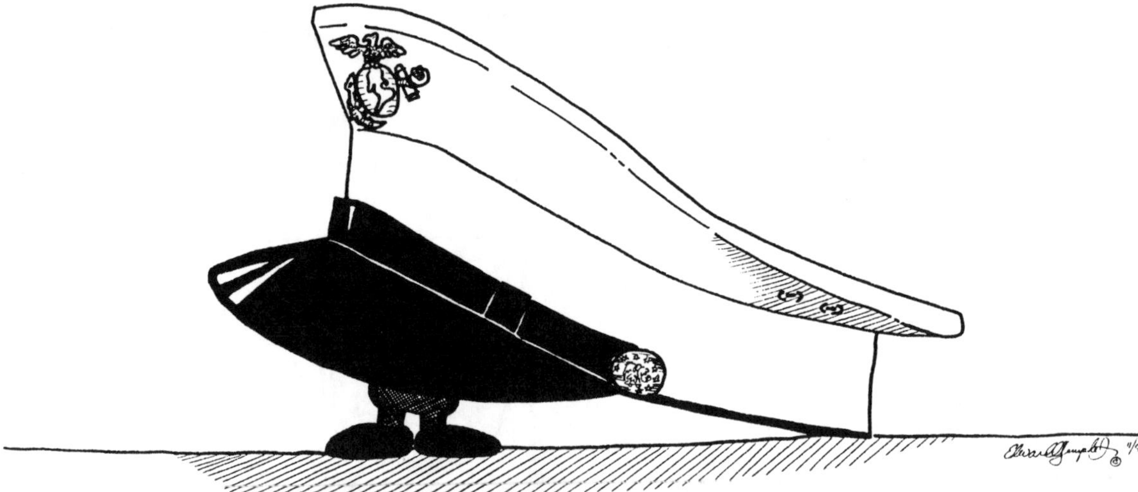

YOU KNOW YOU'RE SHORT WHEN YOUR COVER DOESN'T FIT.

SHORT TIMERS' SWIMMING POOL

MY TURN!!!

EASTER EGG
PAINTING CONTEST
SPLAT
RED
BLUE
BLACK
YELLOW
GREEN
BROW
PUR
BLUE
RED
YELLOW

SHORT TIMER TRAP

SWIM FASTER.

FLU SHOTS

RIFLE REQUAL

NIGHT AT THE COCK FIGHTS.

LEADERSHIP
1ST BN
A. CO
1972
1985
MOTIVATION
ME & CHESTY
SHORT-TIMER KILLER
SGT. MAJOR
LIFER
INK
BLACK
STAMP PAD
STAMP PAD
5-92

21
SPLAT
S.T.
00
©
1-92

$1.00
BALLOON
5-92

USMC
0795
PLAYER 1
PLAYER 2
17 75
US MC
MARINE CORPS
100 PTS
10 PTS
15 PTS
150 PTS
175 PTS
100 PTS
.25
25¢
CORPS BALL
1-92
©

SHORT TIMER MOVIE THEATER

SHORT TIMERS INSPECTION

SERGEANT MAJORS' CAR

M-60's
FIRST TO FIGHT
VICTORY AT HIGH TIDE
U.S.S. NEW JERSEY
TACTICS
THE CENTURIONS
THE WALL
WARFARE
GEARING DOWN
BATTLE LEADERSHIP
SEA POWER
THE VILLAGE
COMBAT
BOATS
PLANES
RIOTS
ATTACKS
WAR IN THE SHADOWS
LIFERS
FIELDS OF FIRE
ACTS OF WAR
SOLDIER'S LOAD
1775 to 1975
FLY HIGHER
CORPS
FIX BAYONETS
INFANTRY IN BATTLE
MRE COOK BOOK II
MRE COOK BOOK III
MRE COOK BOOK IV
SNIPER
ROMMEL
McAURTHER
HOW TO P.T.
RIFLEMAN DODD
BATTLE FOR HUE
MEN AGAINST FIRE
E.S.T.
JAWS
FISHING
MARINE
CAR TIRES
LASER KILL
3rd WORLD
DEVIL DOGS
FORGOTTEN SOLDIER
MARCH TO GLORY
M-1
TANKS
PATTON
LEATHERNECK
M1-A1
MESSAGE TO GARCIA
SEMPER FI
EATING ROAD KILLS
USS IOWA
BATTLE CRY
DIE HAPPY
KEEP KILLIN'
GOV'T CUT-BACK
KILLER ANGELS
CRUSHER
RAT PATROL
WITH THE OLD BREED
COMMON SENSE TRAINING
THE FACE OF BATTLE
Robin Hood
1ST FSSG
COMMAND IN WAR
COMPANY COMMANDER
DESERT
HOW TO EAT SNAKES
CLIMBING
HOW TO EAT GOATS
HOW TO EAT SHEEP
Book Marks
Book Marks
DEATH
STOMPER
GUNG HO
KILL 4 FUN
WOLF PACKS
U-BOATS
SQUADS
PT-109
ONCE A MARINE
BATTLE STUDIES
WESTERN FRONT
TORTURE
SHOCKING
RED BADGE OF COURAGE
BIBLE
LEADERSHIP
AMTRACKS
IWO JIMA
KILL THEM First
GULF WAR
GOLF WAR
©
5-92

NO, UP TO HERE.

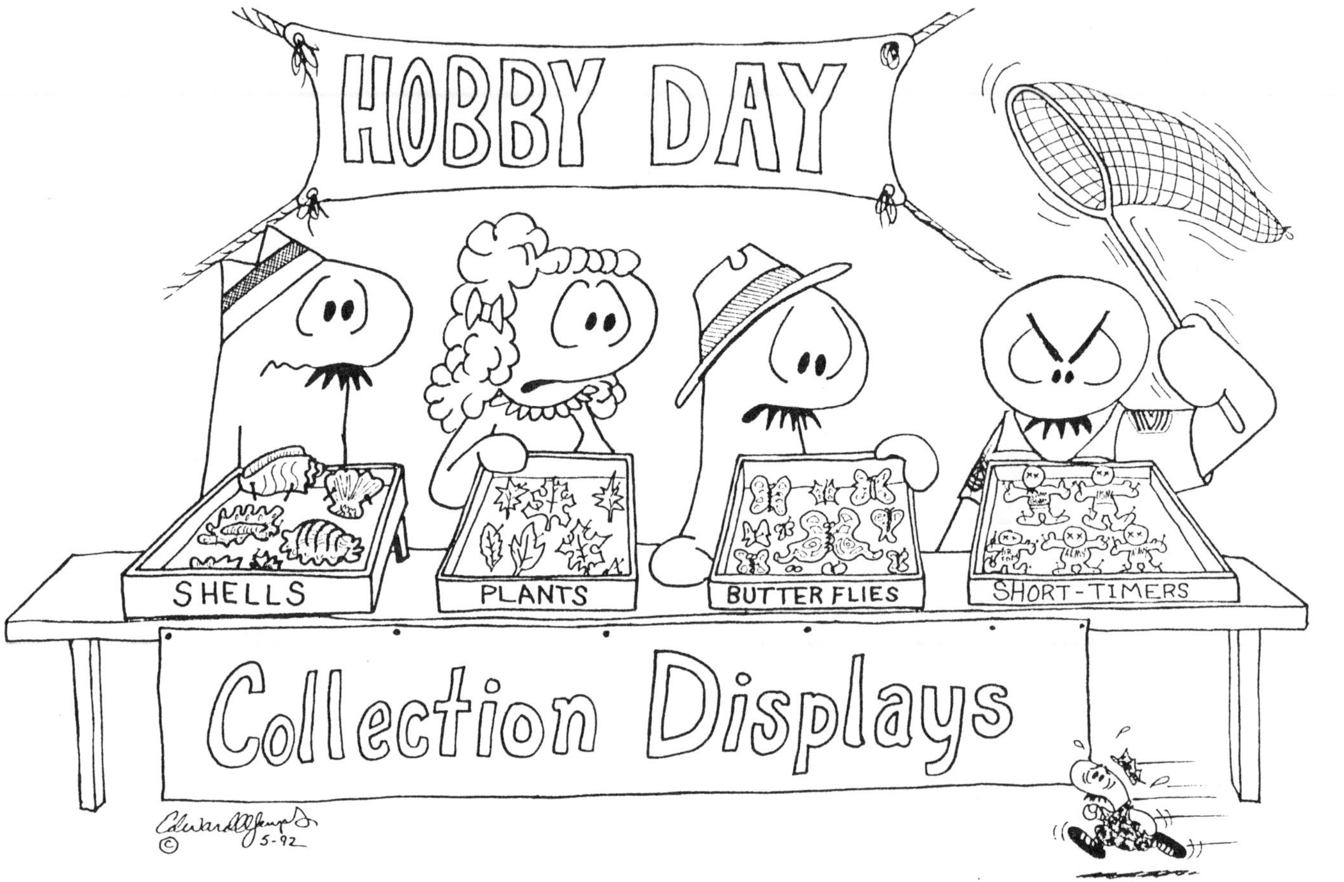
HOBBY DAY
SHELLS
PLANTS
BUTTER FLIES
SHORT-TIMERS
Collection Displays
5-92

SOMEDAY SON, THIS WILL BE ALL YOURS.

IT SAYS," IF YOU HAVE TO BEND OVER TO READ THIS NOTE, YOU'RE A BOOT"

A SHORT TIMER SHOVELING SNOW.

Dear Mom,
How are you? I am fine. Four
days to go. I wi
Edward Young Jr.
©
5-92

HELIUM
© 5-92

IS THAT KNOT SECURED?

MEET PETE. PETE MEET MEAT. EAT MEAT PETE.

MAKE A WISH.

HEADS, YOUR BALL.

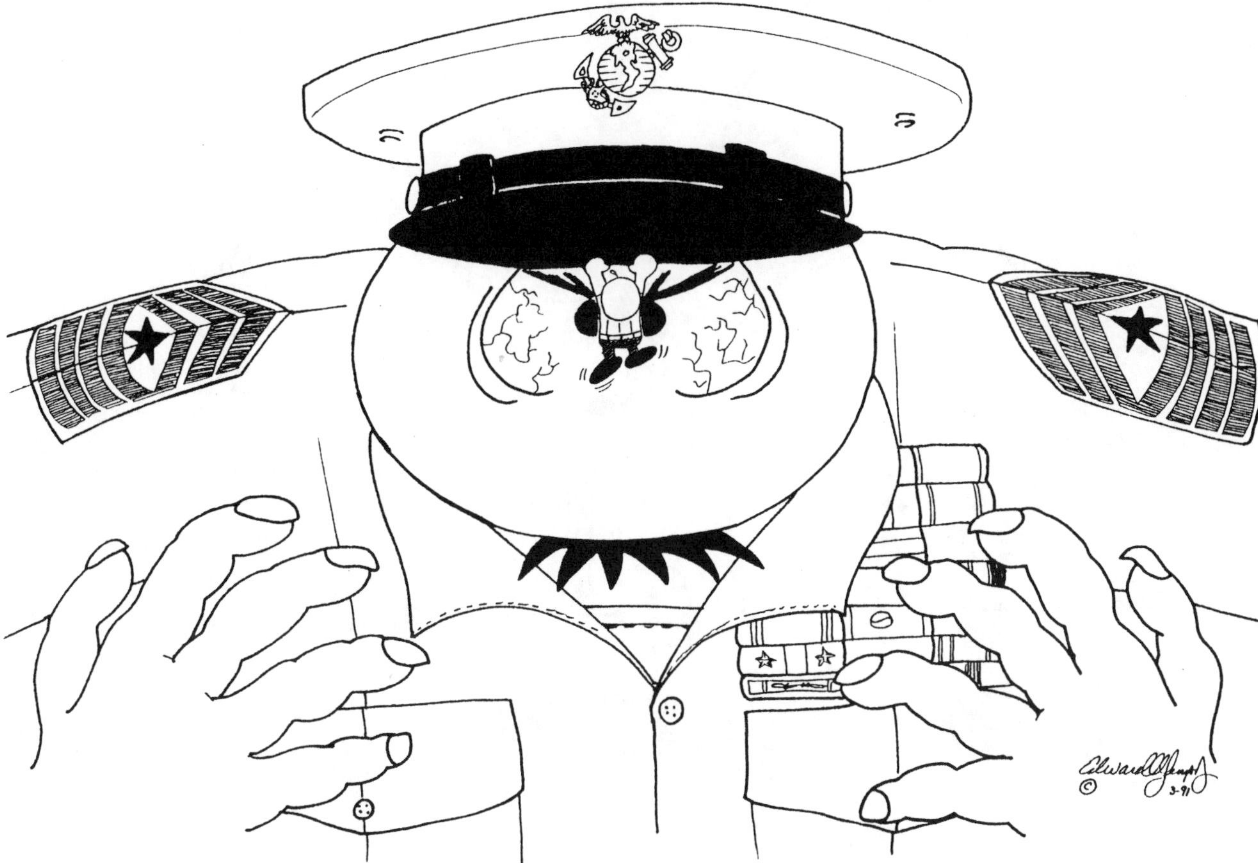

HAND TO HAND COMBAT

ARE YOU ALMOST FINISHED ?

R.R.
77
ACC.
TRACK
POWER
10-91

I WANT TO SEE THIS MOVIE TWO TIMES.

Shhhh... I THINK I HEAR THE SERGEANT MAJOR COMING.

HEY, WHERE'S HARRY ?

INNING 1
BALL 0
STRIKE 0
RBI 0
BATTER 0 1
1 2 3 4 5 6 7 8 9 TOTAL
LIFERS 92 0 0 0 0 0 0 0 0 92
SHORTIES 0 0 0 0 0 0 0 0 0 0
381
347
USMC

CHOW CALL

SPLAING
UNFAIR DAY
Sgt. MAJ.
MGy Sgt.
1st Sgt
MSgt
GySgt
SSGT
SGT
CPL.
Pfc
PVT.
WOW!
18 in a row!!!
© 1-92

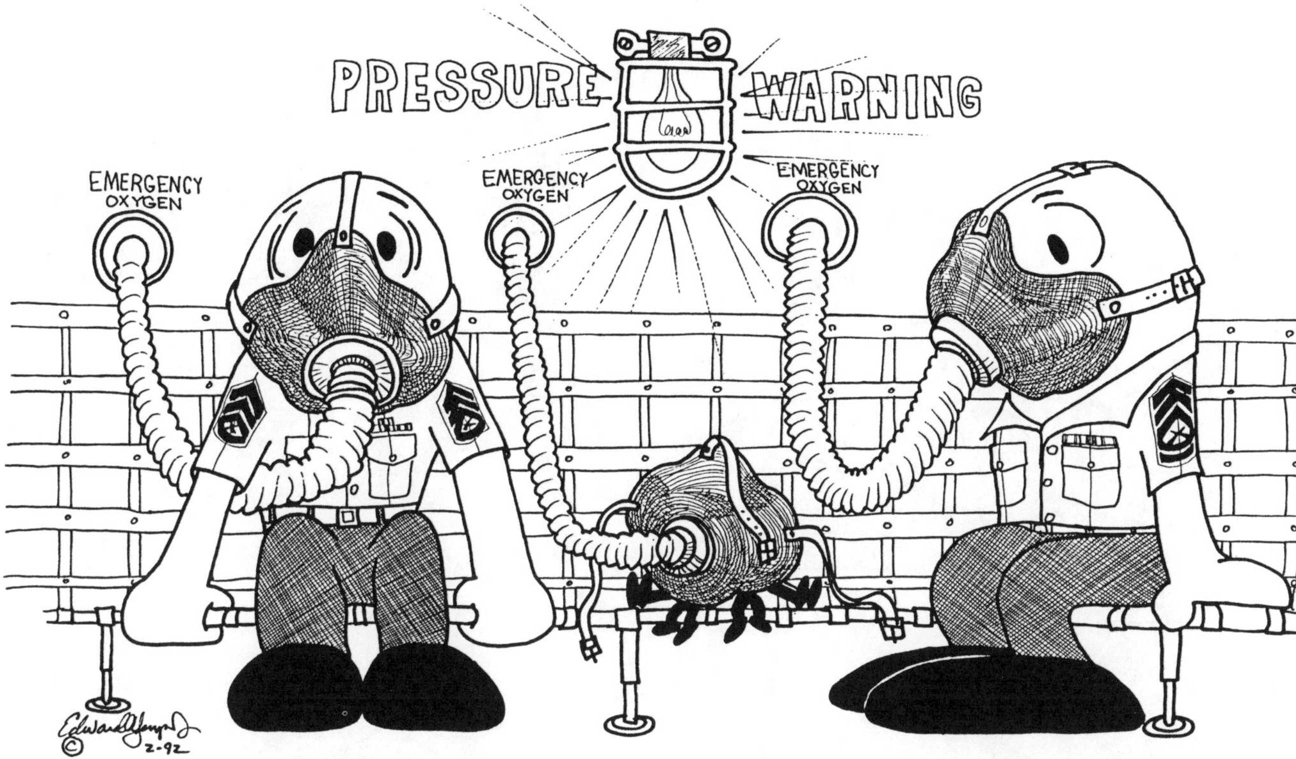
PRESSURE WARNING
EMERGENCY OXYGEN
EMERGENCY OXYGEN
EMERGENCY OXYGEN
2-92

SPLAT
INK
STAMP
ART
© 1-92

SERGEANT MAJOR'S NIGHTMARE

HAPPY SAINT PATRICKS DAY, SERGEANT MAJOR.

AUGMENT FAIRY

DANGER OF DRINKING ALONE

TRICK OR TREAT
TRICK OR TREAT
TRICK OR TREAT
TRICK OR TREAT
© 5-92

YUP, JUST WHAT I THOUGHT ; SHORT TIMERS DISEASE.

THERE REALLY IS A SANTA!

SIR, ALL I SAID WAS,"SOMEDAY YOU'LL BE SHORT TOO SERGEANT MAJOR", AND THEN IT HIT HIM !

MAY
DD-214
HONORABLE DISCHARGE
NAME CPL. HARRY WHO
SSN 177-25-0007
COMPLETED: 4 years of
ACTIVE DUTY
U.S.M.C.
EFFECTIVE: 1600
MAY 12, 1993
SIGNATURE
Commanding officer
© 5-92
5-92

I MADE IT!!!